LITTLE GENIUS, BIG SUMS

MULTIPLICATION

AND

DIVISION

ACTIVITY BOOK

MOONSTONE

Published in Moonstone
by Rupa Publications India Pvt. Ltd 2025
7/16, Ansari Road, Daryaganj
New Delhi 110002

Sales centres:
Bengaluru Chennai
Hyderabad Jaipur Kathmandu
Kolkata Mumbai Prayagraj

P-ISBN: 978-93-6156-775-9
E-ISBN: 978-93-6156-889-3

First impression 2025

10 9 8 7 6 5 4 3 2 1

CONTENTS

CONTENTS

1

INTRODUCTION TO MULTIPLICATION AND DIVISION

MULTIPLICATION AND DIVISION
INTRODUCTION

Hello, math whizzes! Welcome to the marvellous world of multiplication and division! Have you ever wondered how to quickly figure out how many crayons you have in packs or how to share a cake equally with friends? Get ready, because today we're diving into two magical math tricks that help you do just that!

With **multiplication**, we're going to zoom into the fast lane of adding the same number again and again. Imagine you have 3 bowls, and each bowl has 2 apples. Instead of counting them one by one, you can simply multiply: 3 bowls × 2 apples = 6 apples! Isn't that a superpower?

And then, there's **division**—the art of sharing equally! If you have 12 cookies and 3 friends, how many cookies does each friend get? Division breaks it down for us: 12 cookies ÷ 3 friends = 4 cookies each. Sharing made simple!

By the end of this adventure, you'll know how to multiply and divide like a math superhero. So, put on your thinking caps, and let's unlock the secrets of these amazing number tricks! Ready, set, multiply and divide!

VISUAL REPRESENTATION OF
MULTIPLICATION

Multiplication is when we **add the same number many times** or **group things together**. It's like a **shortcut for repeated addition**!

WHAT DOES THAT MEAN?

Multiplication is just a faster way to add the same number again and again.

For example:

- If you have 3 bags with 4 chocolates in each bag, how many chocolates do you have in total?

- You could add: $4 + 4 + 4 = 12$

- Or you could multiply: $3 \times 4 = 12$

Both ways give the same answer: **12 chocolates!**

Multiplication as Repeated Addition

3×2 means adding **2 three times**:

$$2 + 2 + 2 = 6$$

Example:

$$5 \times 2 = 2 + 2 + 2 + 2 + 2 = 10$$

Here's a fun way to think about it:
Multiplication is like building towers. The more groups you add, the taller your tower becomes.

VISUAL REPRESENTATION OF
DIVISION

Let's start with division—a brilliant way to share or split things into equal groups.

WHAT DOES THAT MEAN?

Imagine this: You have **12 delicious cookies**, and you want to share them equally with **3 friends** (including yourself). How many cookies does each person get?

Here's how we figure it out:

- Start with **12 cookies**.

- Divide them equally into **3 groups**.

- Count the cookies in one group: **1, 2, 3, 4**.

So, **12 ÷ 3 = 4 cookies per person**. Easy, isn't it? Division helps us share fairly and figure out equal parts.

Think of division as the **art of splitting things evenly**. Whether it's sharing candies, dividing toys, or grouping pencils, division ensures everything is balanced and fair.

PICTURE EXAMPLE:

Take a look at the set of flowers below.

1. Count the total number of flowers.

2. Split them into **3 equal groups**.

3. Count how many flowers are in each group.

Wow! Isn't it amazing how division organizes things so neatly?

PRACTICE PROBLEMS

Now that you understand how to do multiplication and division, let's work on some simple exercises!

Multiplication:

You have **3 bowls**, and each bowl contains **4 candies**. How many candies do you have in total?

$$3 \times 4 = \underline{\hspace{2cm}} \text{ candies}$$

Division:

You have **12 cookies**, and you want to share them equally among **3 friends**. How many cookies does each friend get?

$$12 \div 3 = \underline{\hspace{2cm}} \text{ cookies per friend}$$

Math becomes more exciting when you practice!

2

BASIC
MULTIPLICATION

Let's kick off with the simplest form of multiplication—
single-digit multiplication! This means multiplying
numbers that are less than 10. You'll be amazed at how
quick and fun it is to work with these small numbers!

SINGLE–DIGIT MULTIPLICATION

Let's dive into some examples:

2 × 2

Imagine you have **2 baskets**, and each basket holds **2 apples**. How many apples do you have in total?

Answer: _____ **apples**.
When you multiply **2 by 2**, it's like saying 2 added to itself once. The result is **4**!

3 × 1

Picture this: You've got **3 cups**, and each cup has just **1 marble** inside. How many marbles are there altogether?

Answer: _____ **marbles**.
Multiplying **3 by 1** doesn't change the number. It's as if each cup brings its own marble to the count!

4 × 3

You have **4 boxes**, and each box is filled with **3 chocolates**. How many chocolates do you have?
Answer: _____ **chocolates**.

Multiplying **4 by 3** is like saying you've got **4 groups** of **3 chocolates**, which makes **12 in total**.

Multiplication is all about **groups of equal size**. Think of it as stacking numbers on top of each other to see how tall the tower gets. The more you practice, the easier it becomes to calculate quickly.

Keep multiplying, and you'll soon be a pro at tackling bigger numbers!

BASIC MULTIPLICATION
WORD PROBLEMS

Now that you understand multiplication, let's explore some word problems. These are fun scenarios that show how multiplication helps us in daily situations!

Example 1

You have **3 baskets**, and each basket contains **4 apples**. How many apples do you have in total?

3 baskets × 4 apples = 12 apples

Example 2

There are **5 shelves**, and each shelf has **2 books**. How many books are there altogether?

5 shelves × 2 books = _____ books

Example 3

Your friend has **6 packs of crayons**, and each pack contains **3 crayons**. How many crayons are there in total?

6 packs × 3 crayons = _____ crayons

Why Word Problems Matter:

Word problems help us see multiplication in everyday situations, like counting apples, arranging books, or sharing crayons. Whenever you hear a number story, think about how you can use multiplication to find the answer.

BASIC MULTIPLICATION
WITH PICTURES

Let's make multiplication even more exciting with some interesting pictures! Look at the images below, count the groups of objects, and then multiply to find the total.

Example 1

2 groups of 3 teddy bears = 6 teddy bears

Example 2

3 groups of 3 apples = _____ apples

Example 3

4 groups of 3 blocks = _____ blocks

Pictures make multiplication fun and easy. Simply count how many objects are in one group and multiply by the number of groups. Keep practicing, and multiplication will become second nature!

BASIC MULTIPLICATION
PRACTICE SET

Now it's time to put your multiplication skills to the test! Solve these problems on your own and write your answers below each question.

Let's start with these:

1. 2×3 = _____

2. 8×2 = _____

3. 5×1 = _____

4. 6×2 = _____

5. 3×3 = _____

6. 2×4 = _____

You're doing fantastic! Let's try a few more:

7. 7×1 = _____

8. 3×5 = _____

9. 8×0 = _____

10. 4×4 = _____

3

BASIC
DIVISION

Now that we've mastered multiplication, let's explore **division**! Division is all about **sharing** or **splitting** things into equal parts to see how much each group gets. It's just as simple as multiplication, but instead of building up, we break things down into smaller, fair shares!

Let's get started and see how division works!

SINGLE–DIGIT
DIVISION

Let's begin with some easy division questions:

$$6 \div 2$$

You have **6 chocolates**, and you want to share them equally with **2 friends**. How many chocolates does each friend get?

Let's divide:
6 chocolates ÷ 2 friends = 3 chocolates per friend

$$8 \div 4$$

Imagine you have **8 pencils**, and you divide them into **4 equal groups**. How many pencils are in each group?

Let's divide:
8 pencils ÷ 4 groups = _____ pencils per group

$$9 \div 3$$

You have **9 cards**, and you decide to share them equally among **3 friends**. How many cards does each friend get?

Let's divide:
9 cards ÷ 3 friends = _____ cards per friend

Division helps us figure out how to split or share things fairly.

Just remember, when you divide, think of breaking numbers into equal parts. It's simple and fun!

BASIC DIVISION
WORD PROBLEMS

Let's solve some engaging division stories! These scenarios will show how division is useful in everyday life.

Scenario 1

You have **12 chocolates**, and you want to divide them equally among **4 friends**. How many chocolates will each friend receive?

12 chocolates ÷ 4 friends = 3 chocolates per friend

Scenario 2

There are **20 apples** to be placed into **5 baskets** so that each basket has the same number of apples. How many apples will go in each basket?

20 apples ÷ 5 baskets = _____ apples per basket

Scenario 3

Your teacher has **15 markers** and splits them evenly among **3 students**. How many markers does each student get?

15 markers ÷ 3 students = _____ markers per student

Division word problems allow us to see how sharing, grouping, and splitting work in real life. If ever you're faced with a situation where something is being divided, use division to uncover the answer!

BASIC DIVISION WITH
PICTURES

Let's make division more exciting with the help of some fun illustrations! Look at the pictures, count the total number of objects, and then divide them into equal groups to see how many are in each group.

Example 1

Divide 4 footballs into 2 groups. How many footballs are in each group?

4 footballs ÷ 2 groups = _____ footballs per group

Example 2

Share 6 doughnuts equally into 3 plates. How many doughnuts does each plate get?

6 doughnuts ÷ 3 plates = _____ doughnuts per plate

Example 3

Share 8 balloons equally among 4 children. How many balloons will each child receive?

8 balloons ÷ 4 children = _____ balloons per child

Using pictures makes division easier to understand. Count the total, divide into groups, and discover how many fit in each group. Keep practicing, and division will soon become second nature!

BASIC DIVISION
PRACTICE SET

Now it's your turn to practice dividing with these enjoyable exercises!
Solve each problem and write your answers in the blanks.

Let's start simple:

1. $6 \div 2 =$ _____

2. $8 \div 4 =$ _____

3. $9 \div 3 =$ _____

4. $12 \div 6 =$ _____

5. $10 \div 5 =$ _____

6. $15 \div 3 =$ _____

Feeling confident? Try a few more:

7. $16 \div 4 =$ _____

8. $18 \div 6 =$ _____

9. $20 \div 5 =$ _____

10. $24 \div 8 =$ _____

You're doing fantastic! Keep solving these problems, and division will
soon feel like second nature. The more you practice, the better you'll
get!

4

MULTIPLICATION WITH

NUMBERS UP TO 10

Now, let's practice multiplying numbers up to 10! We'll use slightly larger numbers and learn how to find the total when we group them together. Multiplication with numbers up to 10 is a great way to strengthen your math skills and make everyday calculations a breeze. Let's get started!

Let's start with some examples:

2×4

Imagine you have **2 baskets**, and each basket has **4 apples**. How many apples are there in total? Let's count: **4, 8**.
So, $2 \times 4 = 8$ apples!

3×3

You have **3 plates**, and each plate has **3 biscuits**. How many biscuits are there altogether? Let's count: **3, 6, 9**.
The answer is **9 biscuits**!

5×2

You start with **5 boxes**, and each box has **2 pencils**. How many pencils are there in total? Let's count: **2, 4, 6, 8, 10**.
So, $5 \times 2 = 10$ pencils!

More Examples to Try:

1. $4 \times 2 = $ _____

2. $3 \times 5 = $ _____

3. $6 \times 1 = $ _____

4. $2 \times 3 = $ _____

5. $7 \times 2 = $ _____

Keep practising, and soon multiplication will be as easy as adding numbers. Happy multiplying!

Fun Activities: Multiply and Explore!

Here are some fun multiplication activities to try. Get ready to draw, count, and multiply while having a blast!

Activity 1

Draw **4 rows of circles**, with **3 circles in each row**.
Now, count how many circles you've drawn in total.

Answer: 4 × 3 = _____ circles

Activity 2

Colour **5 groups of stars**, with **2 stars in each group**.
How many stars are coloured in total?

Answer: 5 × 2 = _____ stars

Why it's fun:

Multiplying larger numbers is exciting because you can quickly find the total without adding them one by one.

MULTIPLICATION WITH NUMBERS UP TO 10:
WORD PROBLEMS

Let's dive into some word problems to practice multiplying numbers up to 10! These stories will help you see how multiplication is used in real-life situations.

Problem 1

You have **4 plates**, and each plate has **3 cookies**. How many cookies do you have in total?

4 plates × 3 cookies = 12 cookies

Problem 2

There are **5 chairs**, and each chair has **2 legs**. How many legs are there altogether?

5 chairs × 2 legs = _____ legs

Problem 3

You have **3 baskets**, and each basket contains **5 oranges**. How many oranges do you have in total?

3 baskets × 5 oranges = _____ oranges

Problem 4

Your teacher brings **6 boxes**, and each box contains **4 pencils**. How many pencils are there in total?

6 boxes × 4 pencils = _____ pencils

Multiplication word problems are a great way to understand grouping and totals. Keep solving these, and you'll soon find multiplication easy and fun!

MULTIPLICATION WORD PROBLEMS WITH
NUMBERS UP TO 10

Let's tackle some multiplication word problems! These will help you see how to solve real-life scenarios by multiplying.

Problem 1

You have **4 baskets**, and each basket holds **2 books**. How many books are there altogether?

4 baskets × 2 books = _____ books

Problem 2

There are **3 tables**, and each table has **4 chairs** around it. How many chairs are there in total?

3 tables × 4 chairs = _____ chairs

Problem 3

You start with **6 trays**, and each tray holds **3 cupcakes**. How many cupcakes do you have in total?

6 trays × 3 cupcakes = _____ cupcakes

Fun Activity

Draw a picture to represent the multiplication problem.
For example, draw **5 plates** with **3 apples on each plate**. Count them to see how multiplication works!

Word problems and drawing activities make math exciting and practical.

MULTIPLICATION WITH
NUMBERS UP TO 10: WITH PICTURES

Let's make multiplication easier by using pictures! Look at the pictures below, count the groups, and then multiply to find the total.

Example 1

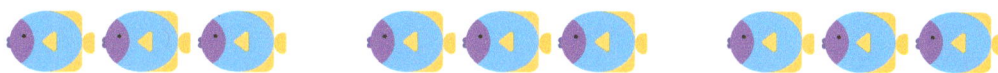

You have **3 groups of 3 fish**. How many fish are there in total?

3 groups × 3 fish = 9 fish

Example 2

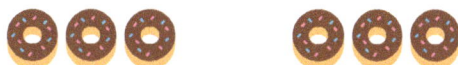

You see **2 plates**, and each plate has **3 doughnuts**. How many doughnuts are there altogether?

2 plates × 3 doughnuts = _____ doughnuts

Example 3

You have **4 groups**, and each group has **3 toy cars**. How many toy cars do you have in total?

4 groups × 3 cars = _____ cars

Pictures help us **see** how multiplication works. Just **count the groups** and **how many items are in each group**, then multiply them to get the total.

Keep practising, and multiplication will become super easy and fun!

PRACTICE EXERCISE

Now it's time to put your multiplication skills to the test with numbers up to 10! Solve each problem and fill in the blanks.

Start with these:

1. $3 \times 3 = $ _____

2. $5 \times 2 = $ _____

3. $3 \times 7 = $ _____

4. $8 \times 2 = $ _____

5. $9 \times 0 = $ _____

6. $4 \times 4 = $ _____

Ready for more? Try these:

7. $3 \times 5 = $ _____

8. $2 \times 6 = $ _____

9. $1 \times 1 = $ _____

10. $5 \times 3 = $ _____

Keep practicing, and before you know it, you'll be a multiplication master! The more you practice, the faster and more confident you'll become.

5

DIVISION WITH

NUMBERS UP TO 10

Division is all about splitting numbers into equal parts to figure out how much goes into each group. Let's explore more examples and problems to deepen your understanding of this concept. Get ready to master division with numbers up to 10!

Let's Try Some Division Examples

Here are some simple examples to help you understand how division works:

Example 1:

$$12 \div 4$$

You have **12 sweets**, and you want to share them equally among **4 friends**. How many sweets does each friend get?

Let's divide: **12 ÷ 4 = 3 sweets per friend**.

Example 2:

$$15 \div 3$$

You start with **15 stickers**, and you share them equally among **3 kids**. How many stickers does each kid receive?

Let's divide: **15 ÷ 3 = 5 stickers per kid**.

Example 3:

$$11 \div 2$$

You have **11 pencils**, and you want to split them evenly into **2 groups**. Each group will get **5 pencils**, but **1 pencil** will be left over.

So, **11 ÷ 2 = 5 remainder 1**.

More Examples to Try:

1.   14 ÷ 2 = _____

2.   8 ÷ 4 = _____

3.   10 ÷ 5 = _____

4.   9 ÷ 3 = _____

5.   16 ÷ 4 = _____

Fun Activities: Divide and Discover!

Let's make division fun with some creative activities. **Draw pictures** to see how groups are made and practise dividing numbers.

Activity 1:

Draw **10 flowers**, then **divide them into 5 equal groups**. How many flowers are in each group?

Answer: 10 ÷ 5 = _____ flowers per group

Activity 2:

Draw **12 cupcakes**, then **split them into 4 groups**. How many cupcakes are in each group?

Answer: 12 ÷ 4 = _____ cupcakes per group

DIVISION WITH NUMBERS UP TO 10:
WORD PROBLEMS

Let's practise division with some **real-life word problems** using numbers up to 10. These short stories will help you learn how to **share or group things equally**!

Problem 1

You have **8 crayons**, and you want to share them equally among **4 friends**. How many crayons does each friend get?

8 crayons ÷ 4 friends = 2 crayons per friend

Problem 2

There are **6 cupcakes**, and you share them equally on **3 plates**. How many cupcakes are on each plate?

6 cupcakes ÷ 3 plates = _____ cupcakes per plate

Problem 3

You have **9 stickers**, and you decide to give them equally to **3 classmates**. How many stickers does each classmate receive?

9 stickers ÷ 3 classmates = _____ stickers per classmate

Problem 4

You have **10 toy cars**, and you want to organise them into **5 boxes**. How many toy cars go in each box?

10 toy cars ÷ 5 boxes = _____ toy cars per box

Why Practise Word Problems?

Word problems show how **division** is used in **everyday situations**. The more you practise, the quicker and better you'll become at solving these problems!

Keep practising, and soon division will feel super easy!

Problem 1

You have **10 toys**, and you want to divide them into **5 equal groups**. How many toys will be in each group?

10 toys ÷ 5 groups = _____ toys per group

Problem 2

There are **8 apples**, and you want to share them equally among **4 friends**. How many apples does each friend get?

8 apples ÷ 4 friends = _____ apples per friend

Problem 3

You have **6 fish**, and you place them into **2 fishbowls** equally. How many fish will be in each fishbowl?

6 fish ÷ 2 fishbowls = _____ fish per fishbowl

Fun Activity:

Come up with your own division scenario and solve it!

Example: Imagine you have **12 balloons** and want to share them with **3 friends**. Draw a picture showing how the balloons are divided equally. Then solve the problem:

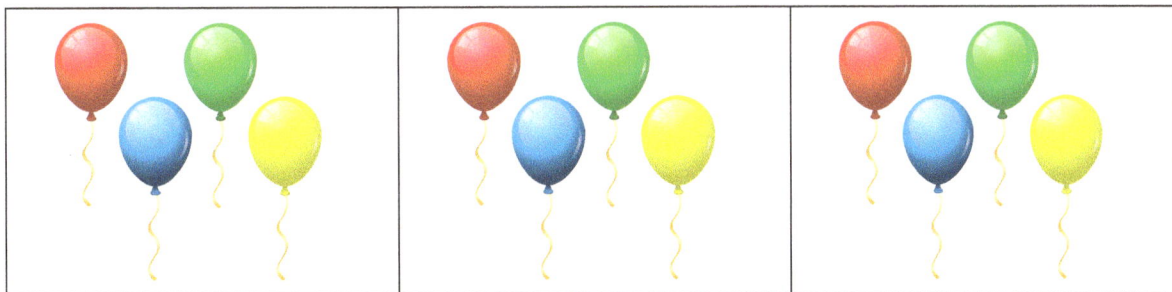

12 balloons ÷ 3 friends = _____ balloons per friend

Practicing division through creative activities makes learning exciting and memorable. Keep up the great work!

DIVISION WITH NUMBERS UP TO 10:
WITH PICTURES

Let's practise **division** using pictures! Look at the pictures, **count how many objects** there are, and then **divide them into equal groups** to see how many are in each group.

Example 1

You have **6 strawberries**, and you want to share them equally into **2 groups**. How many strawberries will be in each group?
6 strawberries ÷ 2 groups = 3 strawberries per group

Example 2

You have **8 balloons**, and you divide them equally into **4 groups**. How many balloons are in each group?
8 balloons ÷ 4 groups = _____ balloons per group

Example 3

You have **6 toy cars**, and you want to share them equally among **3 children**. How many toy cars will each child receive?
6 toy cars ÷ 3 children = _____ toy cars per child

Why Use Pictures?

Division with pictures is a fun way to see how many objects go into each group equally. Just **count the objects** first, then **divide** them to see how many are in each group!

Keep practising, and division will become super easy!

MULTIPLICATION AND DIVISION:

MIXED

Now that we've learned both **multiplication** and **division**, let's **mix them together**! Sometimes, we multiply to find the total, and sometimes we divide to share equally. Let's see how we can solve these problems.

Example 1: Multiplication

$$2 \times 4 = \text{_____}$$

You have **2 plates**, and each plate has **4 cookies**. How many cookies do you have in total?

2 plates × 4 cookies = 8 cookies

Example 2: Division

$$8 \div 2 = \text{_____}$$

You have **8 pencils**, and you share them equally between **2 friends**. How many pencils does each friend get?

8 pencils ÷ 2 friends = 4 pencils per friend

Example 3: Multiplication

$$3 \times 3 = \text{_____}$$

There are **3 baskets**, and each basket has **3 oranges**. How many oranges are there altogether?

3 baskets × 3 oranges = 9 oranges

Example 4: Division

$$12 \div 4 = \text{_____}$$

You have **12 balloons**, and you divide them equally into **4 groups**. How many balloons are in each group?

12 balloons ÷ 4 groups = 3 balloons per group

Why Practise Mixed Problems?

Practising **both multiplication and division** together helps you **understand how they are connected**. Sometimes we need to **multiply to find the total**, and other times we need to **divide to share equally**.

Keep practising, and you'll master both multiplication and division in no time!

Additional Problems: Multiplication and Division

1. $2 \times 3 = $ _____

2. $8 \div 4 = $ _____

3. $5 \times 2 = $ _____

4. $9 \div 3 = $ _____

5. $4 \times 2 = $ _____

6. $12 \div 6 = $ _____

Challenge Problems: Mixed Multiplication and Division

1. $3 \times 2 \div 1 = $ _____

2. $8 \div 2 \times 3 = $ _____

3. $6 \times 2 \div 3 = $ _____

4. $9 \div 3 \times 2 = $ _____

Tip:

When we mix **multiplication** and **division**, remember:

- **Multiply** to find the total when grouping things together.
- **Divide** to share equally or split things into groups.

Keep practising, and you'll master both multiplication and division in no time!

MIXED MULTIPLICATION AND DIVISION
WORD PROBLEMS

Let's solve some fun **mixed word problems**! These stories will help you practise both **multiplication** and **division**.

Problem 1

You have **3 bags**, and each bag has **2 apples**. How many apples do you have in total? Then, you decide to share all the apples equally with **2 friends**. How many apples does each friend get?

- $3 \times 2 = 6$ apples

- $6 \div 2 = 3$ apples per friend

Problem 2

There are **8 chocolates**, and you want to share them equally between **4 friends**. How many chocolates do you have now?

- $8 \div 4 = 2$ chocolates per friend

Problem 3

You have **6 books**, and you organise them in **2 shelves**. How many books are on each shelf? Then, you buy **3 more shelves** and put the same number of books on each shelf. How many books are on each of the new shelves?

- $6 \div 2 = 3$ books per shelf

- $3 \times 3 =$ _____ books per shelf

MIXED MULTIPLICATION AND DIVISION
WITH PICTURES

Let's do some **mixed multiplication and division** using pictures! **Look at the images**, count the objects, and **multiply or divide** as needed.

Example 1

You have **3 groups** of **2 apples** each. Then, you decide to **share them equally** between **2 friends**.

- **3 groups × 2 apples = 6 apples**
- **6 apples ÷ 2 friends = 3 apples per friend**

Example 2

You have **3 plates**, and each plate has **3 cupcakes**. How many cupcakes are there in total?

- **3 plates × 3 cupcakes = _____ cupcakes**

Then, you **divide them equally** among **3 friends**.

- **9 cupcakes ÷ 3 friends = _____ cupcakes per friend**

Example 3

You have **8 balloons**. You **divide them equally** into **4 groups**. How many balloons are in each group?

- **8 balloons ÷ 4 groups = 2 balloons per group**

Then, you **multiply** the number of groups by **2** because you double the balloons for a party!

- **4 groups × 2 balloons = _____ balloons**

Example 4

You have **4 groups of 2 toy cars**. Then, you **share them equally** between **2 friends**.

- **4 groups × 2 cars = 8 cars**
- **8 cars ÷ 2 friends = _____ cars per friend**

Why Use Pictures?

Using pictures helps us **see** how numbers change when we **multiply** or **divide**. Just **count the objects** and follow the steps to multiply or divide!

Keep practising, and soon multiplication and division will feel super easy!

MIXED MULTIPLICATION AND DIVISION
PRACTICE EXERCISE

Now it's time to **practise** what you've learned! Try these **mixed multiplication and division** problems. **Write your answers** below each one.

Let's get started:

1. $2 \times 3 =$ _____

2. $8 \div 4 =$ _____

3. $4 \times 2 =$ _____

4. $9 \div 3 =$ _____

5. $5 \times 2 =$ _____

6. $12 \div 6 =$ _____

Ready for some more mixed problems?

7. $2 \times 4 \div 2 =$ _____

8. $6 \div 3 \times 2 =$ _____

9. $8 \div 2 \times 3 =$ _____

10. $4 \times 3 \div 2 =$ _____

Tip:

- **Multiply** when you are putting equal groups together.
- **Divide** when you are sharing equally or splitting into groups.

You're doing great! Keep practising, and soon you'll be a **master at both multiplication and division**!

7

FUN
ACTIVITIES AND GAMES

MULTIPLICATION TABLE:
FILL IN THE MISSING NUMBERS:

×	1	2	3	4	5	6	7	8	9	10
1		2			5			8		
2		4		8					18	
3	3					18	21		27	
4	4		12				28			
5		10		20		30				50
6				30		42		54		
7		14	21			42		63		
8	8			32		48				
9			27				63			90
10		20				60			90	

SCAVENGER HUNT:
MULTIPLICATION AND DIVISION

It's time for a **Math Scavenger Hunt**! In this fun game, you'll **search for objects** in the room to solve **multiplication and division** problems.

How to Play

- Your **parent or teacher** will say a problem like: **"Find 2 groups of 3 objects"** or **"Find 8 things and share them equally among 4 people"**.

- You have to **find objects** in the room that match the problem. For example, find **2 groups of 3 toys** to solve **2 × 3** or **8 items** to solve **8 ÷ 4**.

- Once you have the objects, **count them** and **say the answer out loud**!

Questions

- **Find 2 groups of 4 apples. How many apples are there in total?**

- **Find 6 pens and share them equally among 3 friends. How many pens does each friend get?**

- **Find 3 groups of 2 books. How many books are there altogether?**

- **Find 12 blocks and divide them equally into 4 groups. How many blocks are in each group?**

Why Play a Math Scavenger Hunt?

A **Math Scavenger Hunt** is a great way to **combine maths with movement and creativity**! It helps you **visualise multiplication and division** and makes learning **active and fun**.

Ready to hunt? Grab some objects and start multiplying and dividing!

CREATE YOUR OWN ACTIVITY:
MULTIPLICATION AND DIVISION PROBLEMS

Now it's **your turn** to create your own **multiplication and division problems**! This activity helps you **practise writing maths problems** for others to solve.

How to Play

- **Think of a fun problem** that uses **multiplication or division**.

- **Write the problem down** and ask your friends or family to solve it!

- You can create **word problems, picture problems**, or just **simple maths problems**.

Examples

- **Word Problem:**
  "I have **3 baskets**, and each basket has **4 oranges**. How many oranges do I have in total?"

- **Picture Problem:**
  Draw **12 balloons**. Now, **divide them into 4 groups**. How many balloons are in each group?

- **Simple Problem:**
  $5 \times 2 = ?$ or $10 \div 2 = ?$

Get Creative!

Once you've created your problems, **let others try them out** and see if they can solve them. You can even make a **booklet of multiplication and division problems** for everyone to enjoy!

Ready to get creative? Grab your pencil and start making some fun maths problems!

43

8

REVIEW AND

ASSESSMENT

Let's **review everything** you've learned so far! Multiplication and division are super important skills that you'll use every day. Here's a quick review of the key concepts.

MULTIPLICATION AND DIVISION
REVIEW OF CONCEPTS

Multiplication

- **Multiplication** is when you **group numbers together** to find the total.

- For example: $3 \times 4 = 12$. You have **3 groups of 4**, which gives you **12 in total**.

Division

- **Division** is when you **split numbers into equal groups** to see how many are in each group.

- For example: $12 \div 4 = 3$. You have **12 items**, and you **divide them into 4 groups**, leaving **3 in each group**.

Things to Remember

- In **multiplication**, the number gets **bigger** because you're grouping items together.

- In **division**, the number gets **smaller** because you're splitting items into equal parts.

Practice both multiplication and division every day to get **faster and better**!

MULTIPLICATION AND DIVISION
PRACTICE TEST

Let's see how much you've learned with this **practice test**.

Solve the problems below and write your answers.

Multiplication Problems:

1. $2 \times 3 =$ _____

2. $4 \times 2 =$ _____

3. $5 \times 1 =$ _____

4. $3 \times 3 =$ _____

5. $6 \times 2 =$ _____

6. $4 \times 4 =$ _____

7. $2 \times 5 =$ _____

8. $3 \times 4 =$ _____

Division Problems:

1. $8 \div 2 =$ _____

2. $9 \div 3 =$ _____

3. $6 \div 2 =$ _____

4. $12 \div 4 =$ _____

WORD PROBLEMS

Word problems are a fantastic way to practise using multiplication and division in everyday situations. Let's explore some fun examples!

1. Sharing Apples

You have 20 apples. You want to share them equally among 4 friends.
How many apples will each friend get?

20 apples ÷ 4 friends = _____ apples per friend

2. Birds on Branches

There are 6 branches, and each branch has 3 birds sitting on it.
How many birds are there in total?

6 branches × 3 birds = _____ birds

3. Pencils for Classmates

You have 24 pencils. You give 6 pencils to each of your 4 classmates.
How many pencils did you give away in total?

6 pencils × 4 classmates = _____ pencils

4. Books on Shelves

You have 5 shelves, and each shelf holds 4 books.
How many books do you have in total?

5 shelves × 4 books = _____ books

CONGRATULATIONS!
You've finished the book and learned all about multiplication and division.
You should be proud of yourself for working so hard!

THIS CERTIFICATE IS AWARDED TO:

~~~~~~~~~~~~~~~~~~~~

FOR SUCCESSFULLY COMPLETING THE

LITTLE GENIUS, BIG SUMS
MULTIPLICATION AND DIVISION BOOK

WELL DONE! YOU'RE ON YOUR WAY TO BECOME A MATH MASTER!

~~~~~~~~~~~~
DATE

~~~~~~~~~~~~
SIGNATURE